Play Smart!

The Science Behind Soccer and Tennis

Discover the science behind every kick, swing, and spin!

by Rudo Kashiri

Curious Minds Explorers Publishing

A Curious Minds Explorers picture book

Ages 8 to 12

Play Smart! The Science Behind Soccer and Tennis

Discover the science behind every kick, swing, and spin!

Illustrations by Roshanay
Published by Curious Minds Explorers Publishing
Stories That Launch Curiosity
www.curiousmindsexplorerspublishing.com

ISBN: 979-8-9996203-3-0

Printed in the United States of America

This book is a work of fiction inspired by real science.
Names, characters, places, and events are used to encourage curiosity and learning and are presented in a fictionalized format.

This Book Belongs To:

DEDICATION

For curious children everywhere, particularly those who delight in movement, play, and asking thoughtful questions.

And for the parents, grandparents, teachers, coaches, and mentors who inspire them to dream boldly, work diligently, and discover all that they are capable of achieving.

Science belongs to everyone.

"We look at science as something very elite, which only a few people can learn. That's just not true. You just have to start early and give kids a foundation. Kids live up, or down, to expectations."

—Mae Jemison, engineer, physician, and astronaut

Meet Nouria

Nouria loves to run, kick, and swing, especially when she's playing soccer or tennis. She's always up for a challenge. But today, she has something new to tackle: a physics science project.

Wait, physics?

Isn't that just for scientists in lab coats?

At first, Nouria can't imagine what science has to do with her favorite games. But with a little help from her wise grandma, Ambuya, and a couple of awesome coaches, she discovers a secret:

Every kick, swing, and spin is powered by science.

Get ready to join Nouria on an exciting adventure where soccer balls curve, tennis balls bounce, and every move is a chance to explore how the world works.

Let's dive in and play smart, because science is hiding in every play.

Nouria burst into the kitchen, nearly juggling her soccer ball. "Ambuya! Guess what?" she blurted. "My science project is about physics in sports!"

Ambuya smiled and set down her teacup with a quiet clink. "Physics, you say? That sounds exciting. What made you choose that topic?"

Nouria grinned, tossing the ball from hand to hand. "Our teacher said we could choose *anything* we're curious about. And I keep wondering this: why does the ball sometimes zoom straight, and other times curve like it has a mind of its own?"

Ambuya nodded. "Ah, that's a wonderful question. Science is everywhere, my dear."

"Even when I'm just playing for fun?" Nouria asked, bouncing the ball once for emphasis.

"Especially then," Ambuya said. "Science is at work every time you kick, bounce, or swing. It's like a secret coach, helping in the background!"

Nouria's eyes widened. "So when I score a goal or smack a tennis ball over the net, I'm doing science?"

Ambuya chuckled. "Absolutely! Every move you make is a tiny experiment. You try something, you see what happens, and you try again, just like a scientist."

My science project is all about physics in sports.
Physics is everywhere.
Motion

"Ready for a little science magic, Nouria?" Ambuya asked. "Physics helps us understand motion, and how things zip, zoom, and spin!"

Nouria giggled. "Like when I blast a soccer ball down the field?"

"Exactly!" Ambuya said. "When you kick the ball, you use force, your superpower push. A bigger force can make the ball speed up more, so a harder kick can send it faster."

Nouria pretended to kick with all her might. "Watch out, world! Incoming mega-kick!"

Ambuya laughed. "And when the ball takes off, it follows a special path through the air. That path is called a trajectory."

Nouria traced a curvy line in the air. "So my soccer ball has its own path every time I kick it?"

"That's right," Ambuya said, drawing a curved arrow on a piece of paper. "Physics helps us figure out why the ball flies high, stays low, goes straight, or curves like it's drawing a rainbow."

Nouria smiled. "I didn't know science could be this much fun."

Ambuya winked. "Coach Erica will help you see these forces on the soccer field!"

Nouria hugged her notebook to her chest. "Let's go. I'm ready to play and investigate!"

Coach Erica jogged onto the field, waving. "Hey, Nouria! Ready for some soccer science?"

Nouria grinned. "Always! What are we learning today, super kicks or mega goals?"

Coach set a soccer ball on the grass. "Soccer is packed with physics, even when you don't notice it. Let's start with force."

She pointed to the ball. "Look, it's not moving. Newton's First Law says it will stay still until something gives it a push."

Nouria tapped the ball with her foot. It rolled forward. "I gave it the push!"

"Exactly," Coach said. "Your foot used force. Newton's Second Law tells us that a bigger force can change an object's motion more, like making the ball speed up faster."

Nouria took a few steps back and kicked the ball harder. It zoomed across the field. "Whoa!" she laughed. "That one had turbo!"

Coach scooped up the ball and kicked it high. "But even a super hard kick can't keep it up forever. The ball always comes back down."

Nouria snapped her fingers. "That's gravity!"

Coach nodded. "Yes. Gravity is the invisible pull that brings the ball back to the ground every time. And the air pushes back a little too," she added. "That's called air resistance, or drag, and it can slow a moving ball."

Nouria nodded. "Got it. Gravity pulls the ball down, and drag can help slow it down."

Whoa! That one went way farther.
Newton's Second Law.
Force (push)
Gravity (pulls down)

Coach bounced the soccer ball and grinned. "Let's play a game. If you can guess where my next kick will go, you win a high-five."

Nouria squared her shoulders. "You're on!"

Coach ran up and kicked the ball with a sneaky spin. It zipped forward, then curved sharply to the side.

"Whoa!" Nouria laughed. "It *swerved*! How did you make it do that?"

Coach winked. "That's spin. When the ball spins, it changes how air moves around it. That sideways push, called the Magnus force, can make the ball curve."

Nouria twirled her hands in the air. "So if I add spin, I can make the ball curve, too?"

"Exactly," Coach said, slapping her a high-five. "Spin can make the ball dance, dip, and dodge, like it's doing a little soccer shuffle."

Nouria grinned. "Next game, I'm bringing my science moves. Get ready."

Coach laughed. "That's playing smart. And it might win you the game!"

Nouria chased after the ball, laughing. "I'm writing this down. Spin is my new secret move!"

"Thank you, Coach! I'm going to practice my spin!" Nouria said, picking up her gear.

The Magnus effect!
Whoa! How did you do that?
The Magnus effect!
Air tends to move faster over the top
→ lower pressure on top
Net force points up (Magnus lift)
Air tends to move slower under the ball
→ higher pressure on bottom

Nouria flopped onto a bench, her notebook balanced on her knees. She flipped it open and started sketching curvy arrows, bouncing balls, and swirling arrows showing spin.

Then she labeled everything like a science detective: Newton's Laws, the Magnus force, and gravity.

"Science really is everywhere on the soccer field," Nouria whispered, like she had discovered a secret.

She held up her notebook for Ambuya to see. "I had no idea soccer had this much science in every kick!"

Ambuya smiled warmly. "And we're not done yet. There's even more science for you to discover."

Nouria zipped her notebook closed. "Okay, what kind of science is in *tennis*?"

Ambuya laughed. "Let's ask Coach Connor."

Nouria sprang up, already imagining tennis balls bouncing like superballs. "I can't wait to meet Coach Connor. Notebook ready!"

She packed her tennis bag with a grin, ready for her next discovery.

Newton's
Laws
Gravity
Magnus
Force

Nouria stepped onto the tennis court, excited. "Hi, Coach!" she said.

Coach Connor waved. "Hi, Nouria! Ready for a little tennis science?"

"I sure am!" Nouria said. "I want to know how science can help my game."

Coach grinned. "Then let's talk about energy, your secret tennis advantage."

"Energy?" Nouria tilted her head. "Like battery power?"

Coach dropped a tennis ball onto his racket. *Boing!* The ball bounced right back up. "When the ball hits the strings, they stretch, then snap back, like a trampoline," Coach said.

Nouria leaned in. "That's awesome!"

"That's elastic energy stored in the strings," Coach explained. "The strings stretch to store energy, then snap back. That energy turns into motion and sends the ball flying."

"So the racket gives the ball extra power?" Nouria asked.

Coach nodded. "Yes, and your body helps, too." He showed her. "Push with your legs, turn your hips, swing your arm, and snap the racket forward."

He swung. **ZOOM!** The ball sailed over the net. "Energy moves like a chain," he said. "It goes from your legs to your body, then down your arm and into the racket."

"Oh! Turn my hips!" Nouria said, turning her body.

Energy your secret tennis superpower!
stored energy
motion
Force
So the racket gives the ball extra power?
Stretched strings store energy.

"Want to learn another tennis secret?" Coach asked.

Nouria nodded eagerly, gripping her racket a little tighter.

"Your racket angle helps control where the ball goes," Coach explained. "If the racket face is too open when you hit the ball, it can pop up and fly long. If it's too closed, it can drive down into the net."

Nouria lined up her shot. "What if I swing faster? Will the ball go farther?"

"Yes!" Coach said. "A faster swing transfers more energy to the ball. Swing faster, and the ball can travel faster and farther. But spin is where the real magic happens."

Coach brushed up on the ball with his racket. "Brush up, and you give the ball topspin. That forward spin makes the ball dip down into the court, so you can hit hard and still keep it in. The Magnus effect helps pull the spinning ball downward, making it drop sooner."

Then he brushed down on the ball. "Brush down, and you make backspin, or a slice. It travels differently through the air and can skid and stay low after it bounces. That can be tricky for your opponent."

Coach tilted his racket to the side and brushed across the ball. "Brush across, and you add sidespin. Sometimes it mixes with topspin. This spin makes the ball swerve left or right in the air."

Nouria watched the ball. "So angle, speed, and spin work together?"

"Exactly," Coach said.

Nouria smiled. She lifted her racket proudly. "Thank you, Coach Connor."

Angle
1
ball flies out
2
ball lands in the court
3
ball hits the net

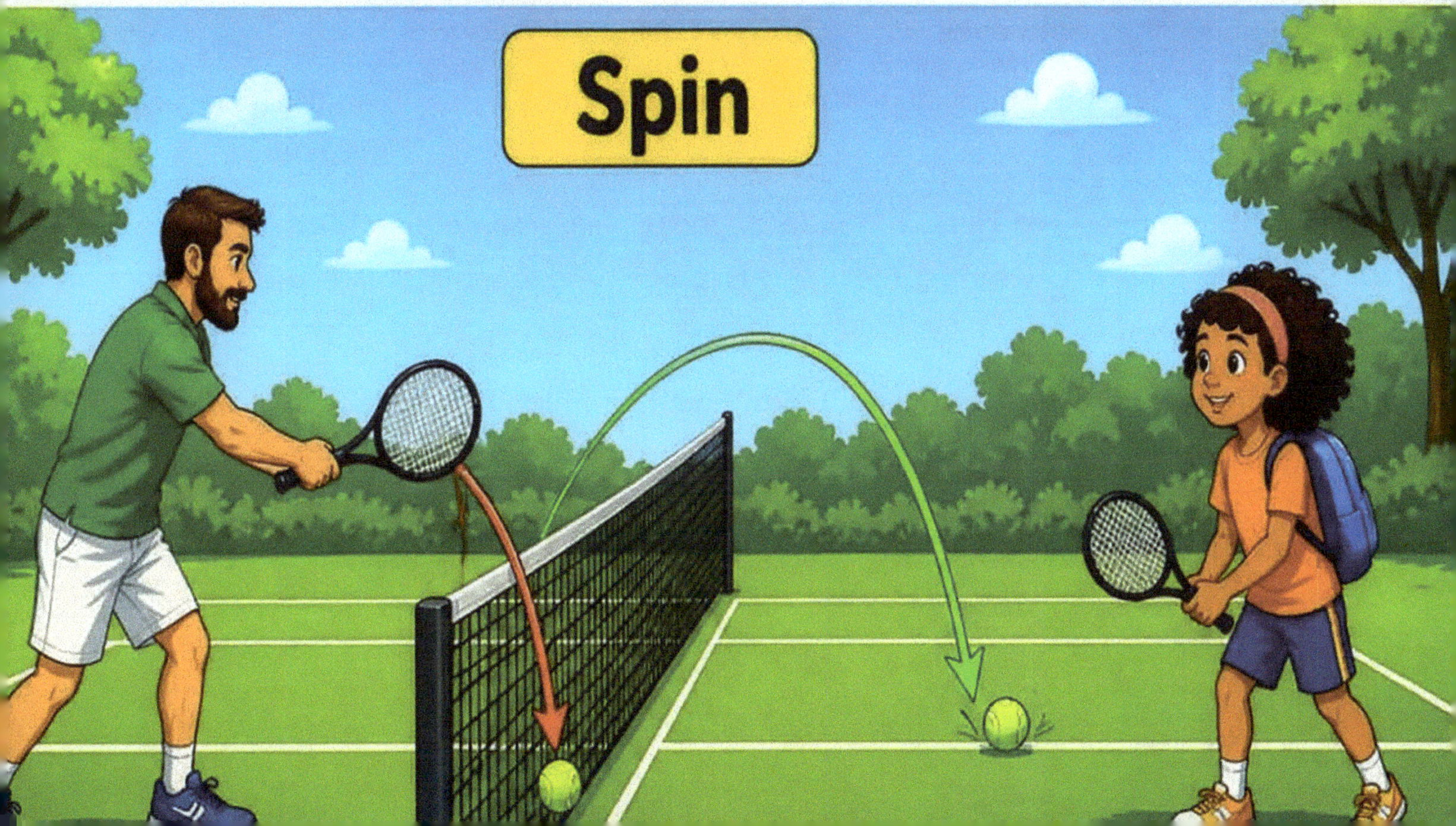
Spin

Nouria skipped along the path, holding Ambuya's hand. Her bag bounced on her shoulder, like it was cheering her on.

"Ambuya, physics is everywhere, every kick, spin, and swing!" Nouria said, practically glowing.

Ambuya smiled. "Tell me what you learned. Start with soccer."

Nouria nodded. "First soccer. If I kick harder, the ball goes faster. Gravity pulls it back down, and spin can make it curve on purpose!"

Nouria kept going. "And in tennis, the strings store energy and bounce the ball back. Even my racket angle can change where it lands."

"I also learned about the Magnus effect in both soccer and tennis. It's amazing how science shows up in more than one sport!" Nouria said with pride.

Ambuya nodded, her eyes twinkling. "That's how science works, my dear. One idea can show up in many places: on a soccer field, on a tennis court, and everywhere you're curious enough to look."

Nouria giggled. "I feel like a real scientist when I ask, 'What if?' and see what happens!"

Ambuya gave her hand a gentle squeeze. "You are a scientist, Nouria. Every time you wonder, test, and explore, you learn just like the greatest scientists."

Nouria looked up, grinning. "I can't wait to finish my project and wow everyone with my discoveries!"

Ambuya's smile grew. "I'm so proud of you," she said.

Nouria sprawled across her desk, colored pencils everywhere. Her notebook was full of sketches: soccer balls zooming, tennis balls bouncing, and curly arrows for spin.

"Let's see. Force, motion, energy, spin," she murmured, drawing a big circle to connect them. "They work together every time I play."

Suddenly, an idea popped into her head. "Time to test it!" Nouria said.

In the hallway, she set up two cones and placed her soccer ball between them.

First, she gave it a gentle kick. It rolled through slowly, like it was tiptoeing. Then she kicked harder. It zoomed past, like it had rocket shoes.

Nouria scribbled in her notebook: *More force = more speed (and more distance).*

Then she tried spin with a tricky little kick. The ball curved around one cone like it was playing tag.

Nouria laughed. "Science isn't just in my notebook. It's in every kick, swing, and bounce!"

Ambuya peeked around the corner and smiled. "That's exactly how a scientist thinks, my dear. Keep testing, keep noticing, and keep asking why."

Gently
Force
Hard!
Whoa! Cool spin!
More force = more speed and distance
Force
Speed
Distance

Nouria stood at the front of the classroom, holding her colorful poster. It was filled with arrows, bouncing balls, and swirling lines.

"Did you know physics shows up in every sport?" she began. "Every kick, swing, and bounce uses science, even when you're just having fun."

Her classmates leaned forward, curious.

"Physics explains why soccer balls curve and tennis balls bounce high," Nouria continued. "It all comes from force, motion, energy, and spin!"

When Nouria finished, her classmates burst into proud applause.

"Well done, Nouria," the teacher said. "You made science come alive."

Nouria beamed. "I couldn't have done it without Ambuya. She showed me that science isn't just in books. It's in every game we play."

Ambuya, sitting in the corner, gave Nouria a proud thumbs-up.

Nouria turned to the class. "Physics is in every sport. What sport do you play?"

The room buzzed as students shared their favorite games and wondered what science they could discover next.

To every curious kid listening, give your curiosity a high-five. Go play, try something new, and see what happens. That's science!

★ SCIENCE IS EVERYWHERE! ★
• Motion
• Force
• Energy
• Spin
Curious
Ask
Think
Learn!
MOTION ★ FORCE
ENERGY
SPIN
Why does my basketball bounce so high?
How does swimming use physics?

STEM Challenge: Be a Science Detective!

Your Mission:

Choose an activity. Change **one** thing. Test it 3 times and note what changes.

1) Choose

- Dancing
- Running
- Any sport/game you love

2) Change ONE thing

- Harder vs. softer
- Faster vs. slower
- Higher vs. lower *or* spin vs. no spin

3) Quick reflection

"I changed ________ and noticed ___________."

Example: "I changed how hard I kicked and the ball went farther."

Observe like a scientist

- Farther or shorter?
- Faster or slower?
- Higher or lower?
- Straight or curved?

What forces are at work?

- **Push** (muscles)
- **Pull** (gravity)
- **Drag** (air resistance)
- **Spin** (can curve)

Safety first ⚠

- Open space
- Soft equipment if possible
- Ask an adult before trying something new

Ambuya's Tip

Results can change; try again, change one thing, and learn.

You are already a scientist.

Science is in your everyday play. **Be a Science Detective!**

Which Forces are at work?

Dance

Basketball

Hockey

Swimming

Running

Baseball

AUTHOR'S NOTE

As a former NASA educator, I have seen the moment it *clicks* for kids, when they realize science is not just something in textbooks. It is part of everyday life. Sports are a perfect way to explore physics because they are full of motion, energy, and exciting surprises.

When you discover that your favorite games are powered by real science, something amazing happens. Your curiosity grows, your confidence builds, and learning becomes a lot more fun.

Want to know a few cool facts? **Physics is AWESOME!**

Physics helps us understand how things move and work, such as:

- **Sports**: Why balls fly, spin, and bounce
- **Games**: How things fall, slide, or crash
- **Inventions**: How cars, planes, and phones work
- **Everyday life**: Why gravity keeps us on Earth

When you know a little physics, you can:

- **Play better** (like curving a soccer ball or adding spin in tennis)
- **Solve puzzles** by figuring out how things work
- **Dream big**, whether you are building cool inventions or asking bold questions

My hope is that this book inspires you to see yourself not just as an athlete, but as a scientist, too.

Remember, **every kick, bounce, and spin is a chance to explore the wonders of physics.**

Rudo Kashiri
Curious Minds Explorers Publishing

www.ingramcontent.com/pod-product-compliance
Lightning Source LLC
LaVergne TN
LVHW070900210926
848482LV00016B/982

* 9 7 9 8 9 9 9 6 2 0 3 3 0 *